Fifty Nifty FACTS ABOUT ELEPHANTS

by J.M. Chapman & S.M. Davis

VULPINE PRESS

Cover and internal design by Vulpine Press LLC

Photo rights obtained from depositphotos
Credited to: Cover – TatianaMakh & vladsilver; back - fouroaks
p. 1 – Andrey_Kuzmin & yobro10; p. 2 – OndrejProsicky; p. 3 – michaklootwijk; p. 4 – JohanSwanepoel
p. 5 – JohanSwanepoel; p. 6 – AOosthuizen; p. 7 – Roman_Baiadin; p. 8 – muha04; p. 9 –jovannig
p. 10 – MKolesnikov; p. 11 – AnkevanWyk; p. 12 – aiisha; p. 13 – mark.markdescande.com
p. 14 – danevski; p. 15 – optikus; p. 16 – cabman237; p. 17 – blackregis2; p. 18 – erbephoto
p. 19 – JohanSwanepoel; p. 20 – fouroaks; p. 21 – fouroaks; p. 22 – abxyz; p. 23 – Gi0572
p. 24 – Gi0572; p. 25 – nevarpp; p. 26 – auntspray, ConchImages, & Chriskruger; p. 27 – Gi0572
p. 28 – smithore; p. 29 – fouroaks; p. 30 – tanyapuntti; p. 31 – JaySi; p. 32 – JHVEPhoto
p. 33 – Nomadsoul1; p. 34 – izanbar; p. 35 – Johnnie_Losev; p. 36 – kung_mangkorn
p. 37 – bondsza; p. 38 –janaka; p. 39 – 2630ben; p. 40 –elvirkin; p. 41 – MattiaATH; p. 42 – Alexis84
p. 43 – impalastock; p. 44 – bruceau; p. 45 – magdalena.paluchowska; p. 46 – povalec
p. 47 – zx6r92; p. 48 – JohanSwanepoel; p. 49 – sgorin; p. 50 – vlasiadis22

Published by Vulpine Press LLC
1400 N 30th Street, Quincy, IL 62301
www.vulpinepress.com
www.fiftyniftyfacts.com

ISBN-10:1-939276-57-8
ISBN-13:978-1-939276-57-5

Asian

African

Elephants are one of two species.

Asian elephants have dented heads, small ears, domed backs, and one finger-like tip on their trunk. African elephants are larger, have domed skulls, concave backs, bigger ears, and two trunk tips.

1

2

Elephants are the largest land mammals.

A typical African bush elephant weighs 13,000 pounds, while an Asian elephant weighs 6,000 pounds. The largest on record was 13 feet tall and weighed around 24,000 pounds.

3

Elephants eat more than 16 hours per day.

Elephants spend up to 75% of their life eating. They consume grass, leaves, small plants and brush, fruit, vegetables, tree bark, and even dig up roots. They don't typically eat peanuts.

4 Elephant herds are led by females.

Living in groups of 8 to 100 individuals, herds comprised mostly of females and their offspring are led by a matriarch. Male elephants leave the herd for good before age 15.

5

Elephant tusks are made of ivory.

The long, curving incisors are made of dentine and an outer layer of tooth enamel. Ivory is highly prized in many cultures, and each year, 30,000 wild elephants are illegally killed for their ivory tusks.

6 Elephants have small eyes.

Their eyesight is not very good, and they rely mostly on their trunk for direction. As their eyes are on the sides of their head, they have good peripheral vision, but poor depth perception.

7

Elephants have six sets of teeth.

They have 12 premolars, 12 molars and 2 tusks. Premolars are used to tear and grind the majority of their foods, while molars work it down to a paste. Tusks are used for digging, display, and defense.

8

Elephants have the longest pregnancy.

Expectant mothers are pregnant for 22 months. This allows for brain and trunk development. Newborns are unstable and may trip over their trunk but can soon keep up with the herd.

9

Elephant trunks have many functions.

Trunks are used to grasp, touch, smell, amplify trumpeting sounds, suck and blow water or dust, and for breathing. A trunk can reach up to 23 feet or lift up to 770 pounds and still perform delicate tasks.

10

Elephant trunks are very muscular.

Within the trunk,150,000 muscle units called fasci-cles combine into 40,000 muscles. In comparison, a human's entire body only has 639 muscles.

11

Elephants are quite tall and heavy.

Over the course of tens of millions of years, they grew in size and weight in order to protect themselves from predators until no carnivore could reliably hunt them.

12

Some elephants are endangered.

There are only 40,000 Asian elephants and 500,000 African elephants left in the wild. The decline is caused by ivory poaching and habitat destruction.

13

Elephants eat 600 lbs. of food/day.

Their bulkiness requires them to intake an enormous amount of nutrients and calories constantly throughout the day. However, their bodies only digest 40% of it.

14

Elephants can live to be 70 y/o.

Because of their long life expectancy, herds can expect to house numerous generations. Captive elephants live even longer, the oldest living to be 86.

15

Elephants can swim long distances.

Using all four legs to paddle and their trunk as a snorkel to breathe while their mouth is underwater, elephants can swim up to six hours and 30 miles in one stretch.

16

Elephants have a slow heartrate.

An adult elephant's heart weighs 26 to 46 pounds and only beats 30 times per minute. It speeds up by 8 to 10 beats per minute when they lay down.

17

Elephants have the largest brain.

Parts of the brain called the limbic system and the hippocampus, important for emotions and memory, are especially large and well-developed in elephants.

18

Elephants have thick skin.

Their skin is up to an inch thick in places, but it's considerably thinner on the ears, around the eye, and on the belly. The wrinkles help cool the body.

19

Elephants can't jump.

Their weight is far too substantial for all four of their legs to leave the ground at once. Their shoulders are also not sturdy enough to reliably handle the force of landing.

20

Elephant babies weigh 230 lbs.

That's about the weight of one and a half full-grown men! Calves depend on their mother for around a year before becoming independent.

21

Elephant teeth grow forward.

Instead of growing up through the gums like most mammals, elephant teeth grow forward from the back of the mouth and push old teeth out the front.

22

Elephants listen with their feet.

Sound wave vibrations carried through the ground from stomps and trumpets are detected through their sensitive feet and used as warning signals.

Elephant ears help cool them.

Blood vessels close to the surface of the skin allow heat to escape, even more so via fanning. Their ears are also as unique as human fingerprints.

23

24

Elephants live in Asia and Africa.

They thrive in regions filled with plentiful trees and sources of water. Their favorite spots include the outskirts of African forests and Asian jungles.

25

Elephant babies act like babies.

Like human babies sucking their thumbs, newborn elephants often suck on the tip of their trunk for comfort. They also huddle for warmth and love.

26

Elephants have strange relatives.

While elephants are most closely related to wooly mammoths, the hairy giants are extinct. Their closest living relatives are hyraxes and manatees.

27

Elephants hear very low pitches.

They can hear sounds as low as 5 hertz; humans only hear down to 20. Low sounds travel enough distance for elephants to communicate over 2.5 miles.

28

Elephant trunks have no bones.

The trunk is a fusion of the elephants upper lip and nose. A unique pair of nerves run along the sides, and cartilage separates the nostrils at the tip.

29

Elephant trunks can hold 2 gallons.

Elephants don't drink through their trunk. Instead, they suck water up to blow into their mouths or shower themselves and their young to keep cool.

30

Elephants are born hairy.

Newborns have thick tufts of hair on their heads and back that falls out as they age. Adult elephants have sparse hair, which cools their skin.

Elephants need mud baths.

31

Because elephants don't have fur to protect them from harsh sun rays, they cover their skin with a layer of mud as a sunscreen to prevent sunburn.

32

Elephants' front legs offer support.

Due to their massive heads and tusks, an elephant's front legs support more of their weight. Asian elephants can be trained to do a handstand.

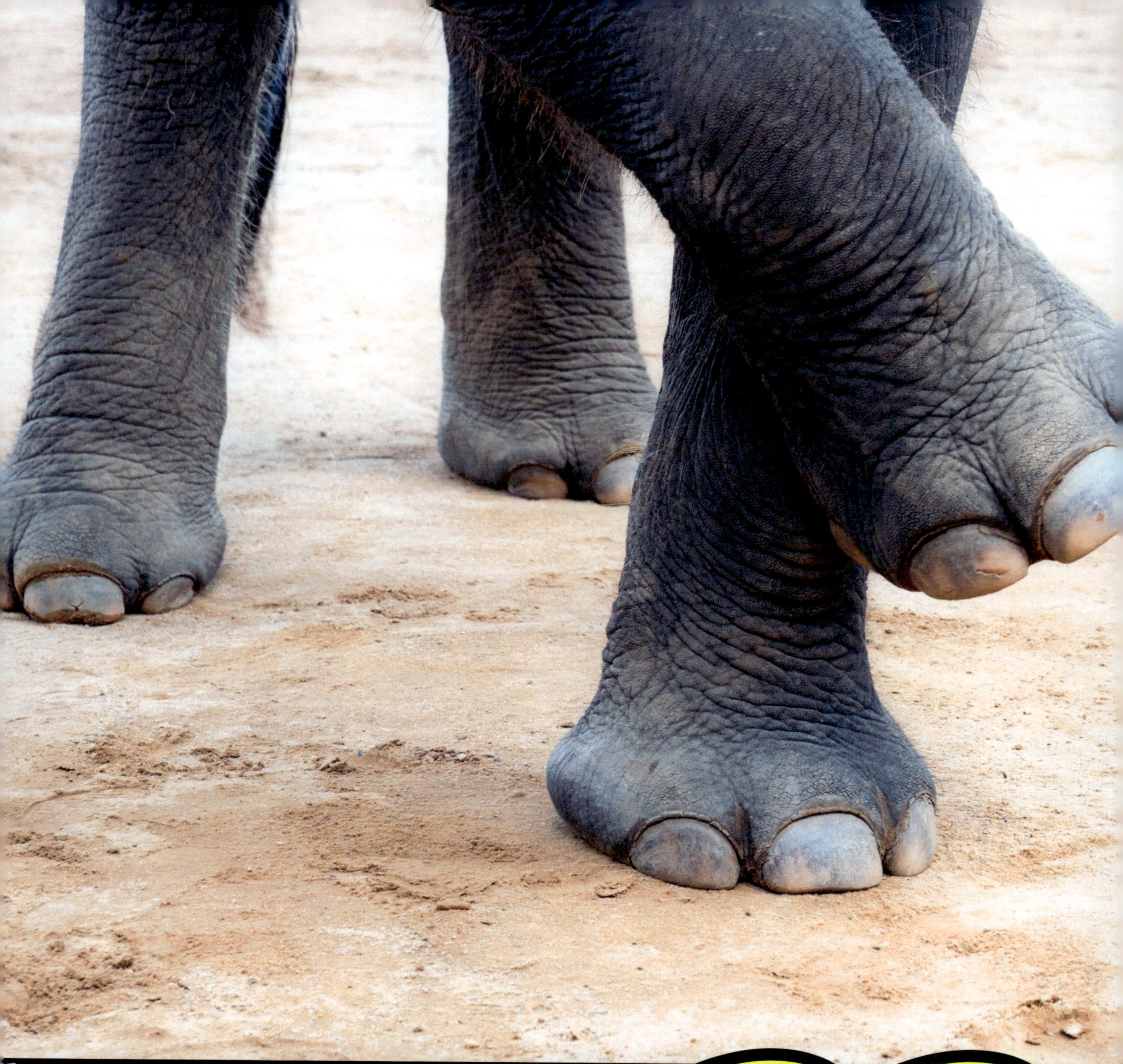

33

Elephants cannot rotate their limbs.

To support their massive weight, elephant leg bones remain fixed and do not rotate. Elephants are the only animals in which all four knees bend backward.

34

Elephants can get tuberculosis.

Diseases are usually specific to one species, but tuberculosis can be passed between humans and elephants, especially in circuses and zoos.

35

Elephants aren't all gray.

After mud baths, they may appear reddish or brown. Rare albino elephants can be white or pink. Aged Asian elephants get pink patches with gray spots.

36

Elephants digest food for a day.

Everything they eat passes through over sixty feet of intestines. Undigested seeds sprout vegetation and can act as a food source for other animals.

37

Elephants' temples secrete fluids.

During musth, male elephants produce up to 60 times more aggression-causing testosterone and secrete an odorous fluid from temporal glands.

38

Elephants care for others' babies.

Mother elephants appoint other elephants to help care for their calves. The offspring are more likely to survive if raised around their wise grandmothers.

39

Elephants rarely get cancer.

Elephants have 20 copies of gene for a cancer blocking enzyme called p53. Humans only have 2 copies of the same tumor suppressor gene.

40

Elephants can work for humans.

Originally plowing farms, Old Bet was the first circus elephant in 1805. They've also been used in forestry to move logs and as transportation.

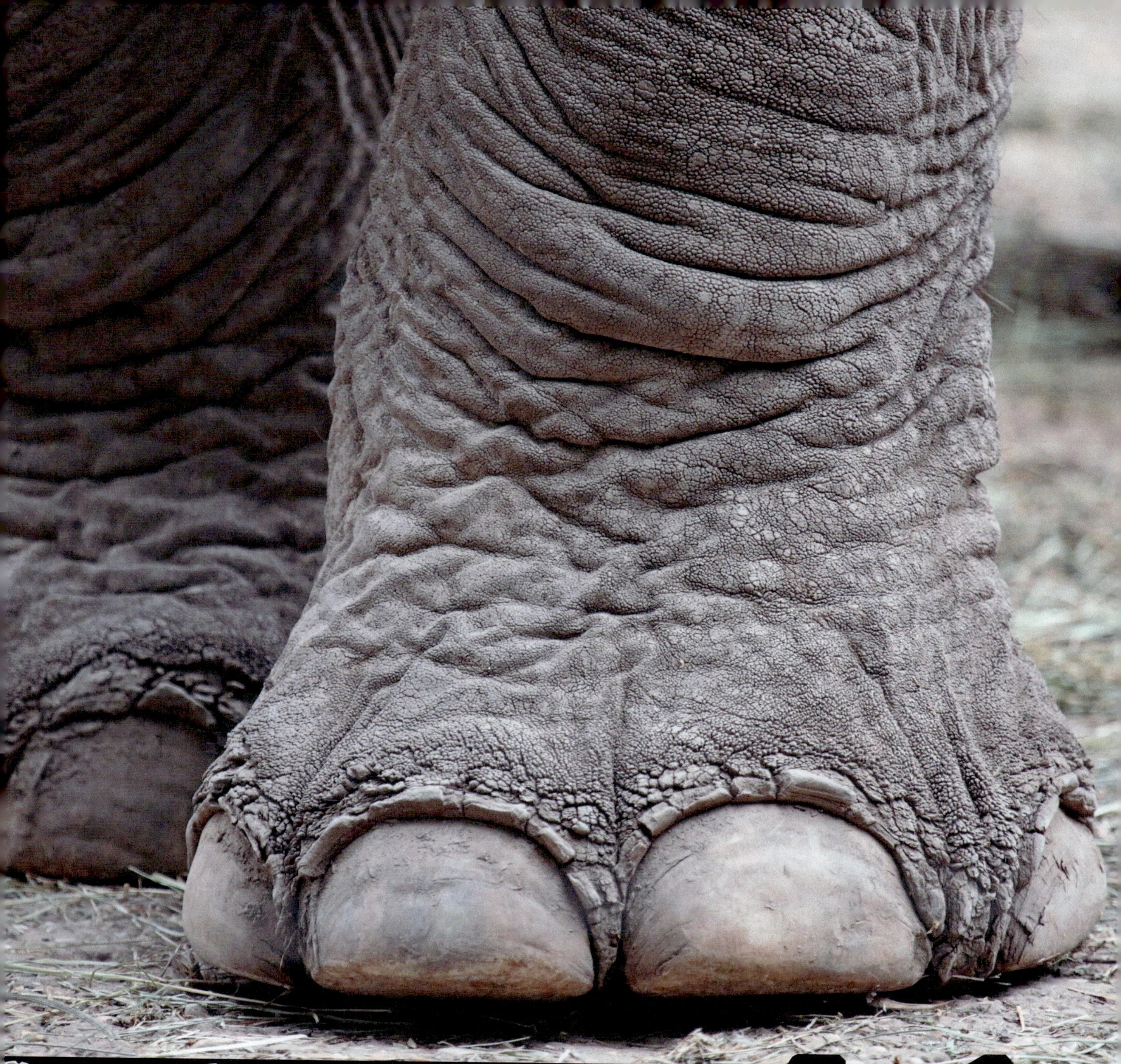

41

Elephants have six toes.

Elephants have five regular toes that point forward and a sixth one in back to support weight. The number of toenails varies - either 5 or 4 nails on front feet and 5, 4 or 3 on back.

42

Elephants hear far-away noises.

Elephants can hear other elephants trumpet 5 miles in the distance. They can also hear a helicopter 90 miles and thunder 150 miles away.

43

Elephants are the best smellers.

With 2000 genes for smell, they can detects scents about four times better than a blood hound can. They can smell a watering hole twelve miles away.

44

Elephants mourn deaths.

Sympathetic in nature, they become distraught when other elephants or human friends pass away, gathering around homes or bones to grieve.

45

Elephants have chins.

Besides humans, only elephants have a bony protuberance on their lower jaw that forms a chin. Their chin anchors their flexible and mobile lower lip.

46

Elephants sleep very little.

While elephants can doze standing up, they must lie down to get deep, REM sleep, but only do so for 2 to 3 hours a day. They dream as well.

47

Elephants drink 50 gallons/day.

They always linger near watering holes to quench their never-ending thirst. Due to lack of sweat glands, they must pee a lot, up to 42.5 gallons in a single go.

48

Elephant tusks weigh a lot.

Together they can weigh up to 440 lbs. Older elephants have larger tusks because they've had longer to grow. Genetics also determines tusk size.

49

Elephants are afraid of bees.

Though their skin is too thick be stung, bees may enter their sensitive trunk. Some farmers in Africa put bee hives near their fields to keep elephants out.

50

Elephants have a master tusk.

Elephants prefer to use either the right or left side of their body and thus use one tusk more often. The dominant tusk shows more wear over time.

Made in the USA
Monee, IL
05 November 2026

67380133R00033